Introduction to Food Waste

Chapter 1: Food Waste in the Curriculum

Chapter 2: Practical Classroom Activities

Chapter 3: Reducing Food Waste in Schools

Chapter 4: Monitoring and Evaluating Food Waste

Chapter 5: Food Waste and Sustainability

Conclusion

Introduction to Food Waste

Food waste refers to food that is discarded or left uneaten at any stage of the food supply chain, from production and processing to retail and consumption. It encompasses a wide range of materials, including uneaten portions of meals, expired products, and food that is spoiled during transportation or storage. The distinction between food loss and food waste is important; food loss occurs during the earlier stages of production, post-harvest, and processing, while food waste typically happens at the retail and consumption stages.

Food waste can occur for various reasons. At the production level, crops may be left unharvested due to economic factors, such as low market prices or labor shortages. During processing, food may be discarded because it does not meet quality standards or due to inefficiencies in the production line. At the retail stage, food can be wasted due to overstocking, improper storage, or damage during transportation. Consumers contribute to food waste by buying more than they need, improper storage at home, and discarding food that has passed its sell-by date but is still safe to eat.

Understanding the multifaceted nature of food waste is crucial for developing effective strategies to mitigate it. By recognizing the various stages at which food waste occurs and the underlying causes, educators can better inform students about the importance of reducing waste and the steps that can be taken at individual and systemic levels.

The Global Scale of the Problem

Food waste is a significant global issue with far-reaching environmental, economic, and social impacts. According to the Food and Agriculture Organization (FAO) of the United Nations, approximately one-third of all food produced for human consumption is lost or wasted globally, amounting to about 1.3 billion tons per year. This staggering amount of waste occurs despite widespread food insecurity and hunger, highlighting the inefficiencies in the global food system.

The environmental impact of food waste is substantial. When food is wasted, all the resources used to produce, process, transport, and store that food are also wasted. This includes water, energy, and land. Additionally, food waste contributes significantly to greenhouse gas emissions. When organic waste decomposes in landfills, it produces methane, a potent greenhouse gas that is far more effective at trapping heat in the atmosphere than carbon dioxide. Reducing food waste, therefore, is a critical component of efforts to combat climate change.

Economically, food waste represents a significant loss of money for farmers, businesses, and consumers. The FAO estimates the economic cost of food waste at around $1 trillion per year. This financial loss is felt across the supply chain, from producers who lose potential income to consumers who waste money on food they do not consume.

Socially, the vast amount of food wasted each year is particularly troubling given the prevalence of hunger and malnutrition around the world. Efficient management and redistribution of surplus food could play a crucial role in alleviating food insecurity. By addressing food waste, we can make progress toward creating a more sustainable, equitable, and efficient global food system.

Causes of Food Waste

Understanding the root causes of food waste is essential for developing effective strategies to mitigate this pressing issue. Food waste occurs at various stages of the food supply chain, from production to consumption, each with its own unique set of challenges and contributing factors.

Agricultural Practices

Agricultural practices are a significant contributor to food waste, with losses occurring during planting, growing, and harvesting. One primary reason is that crops often fail to meet market standards for size, shape, or appearance, leading to their rejection and subsequent waste. For instance, fruits and vegetables that are deemed cosmetically imperfect are frequently left unharvested or discarded, even though they are perfectly edible. Additionally, pests, diseases, and adverse weather conditions can damage crops, resulting in partial or complete loss of harvests.

Another factor is the timing of harvests. Crops harvested too early or too late can suffer from spoilage or reduced quality, rendering them unsellable. Inefficient or outdated harvesting techniques can also lead to substantial losses. Furthermore, the lack of proper storage facilities on farms can cause significant post-harvest losses, as produce may spoil before it can be transported to markets. Addressing these issues requires investment in better farming practices, infrastructure, and technologies to minimize losses at the agricultural stage.

Supply Chain Inefficiencies

Supply chain inefficiencies play a crucial role in food waste, particularly during the processing, transportation, and retail stages. Poor handling and storage practices can cause food to spoil before it reaches consumers. For example, perishable items like dairy products, meat, and fresh produce require specific temperature conditions during transportation and storage. Any deviation from these conditions can lead to spoilage.

Additionally, logistical challenges, such as delays in transportation and improper packaging, can result in significant food losses. In retail settings, overstocking and improper inventory management often lead to products exceeding their sell-by dates and being discarded. To combat these inefficiencies, there is a need for improved logistics, better storage solutions, and more accurate demand forecasting to reduce the amount of food wasted throughout the supply chain.

Consumer Behavior

Consumer behavior is a major factor contributing to food waste, particularly in developed countries. One of the primary issues is over-purchasing, where consumers buy more food than they can consume before it spoils. This is often driven by promotional deals and bulk purchasing incentives that encourage buying in excess. Additionally, improper storage at home, such as not using airtight containers or refrigeration, leads to food spoilage.

Another significant issue is the misunderstanding of date labels. Many consumers confuse "sell-by," "use-by," and "best-before" dates, often discarding food that is still safe to eat. Portion sizes also play a role, with large servings in restaurants and at home resulting in uneaten leftovers that are frequently thrown away. Educating consumers about proper food storage, understanding date labels, and encouraging mindful purchasing and portioning can help reduce food waste at the household level.

Impacts of Food Waste

The impacts of food waste are far-reaching, affecting the environment, economy, and society at large. Understanding these impacts is crucial for recognizing the urgency of addressing food waste and implementing effective solutions.

Environmental Impacts

Food waste has significant environmental consequences. When food is discarded, all the resources used to produce, process, and transport that food are wasted as well. This includes water, energy, land, and labor. For example, agriculture accounts for about 70% of global freshwater use. When food is wasted, the water used to grow that food is also wasted, contributing to water scarcity.

Additionally, food waste contributes to greenhouse gas emissions. When organic waste decomposes in landfills, it produces methane, a potent greenhouse gas that is far more effective at trapping heat in the atmosphere than carbon dioxide. According to the Food and Agriculture Organization (FAO), if food waste were a country, it would be the third-largest emitter of greenhouse gases after the United States and China. Reducing food waste can significantly mitigate these environmental impacts, conserving natural resources and reducing greenhouse gas emissions.

Economic Costs

The economic costs of food waste are substantial. Globally, the FAO estimates that food waste costs about $1 trillion annually. This financial loss affects all stakeholders in the food supply chain, from farmers to consumers. For farmers, wasted food represents lost income and wasted resources. For businesses, it leads to higher costs due to inefficiencies and the need to dispose of unsellable products.

Consumers also bear economic costs through wasted household expenditures. When food is purchased but not consumed, the money spent on that food is effectively wasted. Reducing food waste can lead to significant economic savings for individuals, businesses, and economies as a whole, making it an important target for cost-saving initiatives.

Social Implications

Food waste has profound social implications, particularly in the context of global hunger and food security. Despite the vast amount of food wasted, millions of people around the world suffer from hunger and malnutrition. The FAO estimates that about 820 million people do not have enough to eat. Efficient management and redistribution of surplus food could play a crucial role in addressing this issue.

Furthermore, food waste exacerbates inequalities. In many cases, the regions that produce the most food waste are not the ones facing the highest levels of food insecurity. Addressing food waste can help bridge this gap, ensuring that more food reaches those in need. Socially, reducing food waste fosters a culture of responsibility and sustainability, encouraging communities to value food and resources more highly.

The Role of Education in Food Waste Management

Education plays a pivotal role in addressing food waste. By informing and engaging students, educators can foster a generation that is more conscious of their consumption habits and committed to sustainability.

Importance of Teaching Food Waste

Teaching about food waste is crucial for several reasons. Firstly, it raises awareness among students about the scale and impact of the problem. Many students are unaware of how much food is wasted globally and the consequences of this waste. By integrating food waste education into the curriculum, students can learn about the environmental, economic, and social impacts of food waste, fostering a deeper understanding of the issue.

Secondly, education empowers students to take action. Knowledge about food waste can inspire students to adopt more sustainable behaviors, such as reducing their own food waste, participating in food recovery programs, and advocating for change within their communities. Educators can provide students with the tools and knowledge they need to make informed decisions about food consumption and waste management. Ultimately, teaching about food waste can lead to a more sustainable future, as students carry these lessons into adulthood and influence others.

Educational Goals and Objectives

Educational goals and objectives for food waste management should be clearly defined to ensure effective learning outcomes. One primary goal is to increase students' knowledge about the causes and impacts of food waste. This includes understanding the various stages of the food supply chain where waste occurs, as well as the environmental, economic, and social consequences of food waste.

Another key objective is to develop critical thinking and problem-solving skills. Students should be encouraged to analyze the factors contributing to food waste and to think creatively about solutions. This can be achieved through project-based learning, where students design and implement initiatives to reduce food waste in their schools or communities.

Additionally, fostering a sense of responsibility and stewardship is important. Students should recognize their role in contributing to and reducing food waste. Educational programs should aim to instill values of sustainability and conservation, encouraging students to adopt behaviors that minimize waste and promote efficient use of resources. By setting these educational goals and objectives, educators can create a comprehensive approach to food waste management that prepares students to be informed and active participants in creating a sustainable future.

Chapter 1: Food Waste in the Curriculum

Integrating food waste topics into the curriculum is essential for fostering awareness and action among students. This chapter explores how educators can seamlessly incorporate food waste education across various subjects and grade levels. By embedding these concepts into everyday learning, teachers can equip students with the knowledge and skills to understand and address the issue of food waste, promoting sustainable practices both in and out of the classroom. Through thoughtful lesson planning, age-appropriate learning objectives, and the development of critical thinking skills, educators can make food waste a key component of their educational programs.

Integrating Food Waste Topics

Integrating food waste topics into the curriculum can enrich students' learning experiences while raising awareness about an important global issue. By incorporating food waste education across various subjects, teachers can create a holistic understanding of the problem and inspire students to take meaningful action.

Cross-Curricular Opportunities

One of the most effective ways to integrate food waste topics into the curriculum is through cross-curricular opportunities. This approach allows students to see the interconnectedness of food waste with various academic disciplines and real-world applications.

Science:

In science classes, food waste can be explored through lessons on ecosystems, the carbon cycle, and sustainability. Students can learn about the environmental impacts of food waste, such as greenhouse gas emissions and resource depletion. Hands-on activities, such as composting projects and experiments on decomposition, can provide practical insights into the scientific principles behind food waste management.

Mathematics:

Mathematics offers numerous opportunities to incorporate food waste topics. Teachers can use data on food waste to teach concepts such as statistics, percentages, and graphing. For example, students can analyze data on the amount of food waste generated by their school or community, calculate the financial and environmental costs, and create visual representations of their findings. This not only enhances their mathematical skills but also fosters a deeper understanding of the scale of food waste.

Social Studies:

In social studies, food waste can be examined from historical, economic, and cultural perspectives. Students can study the history of food production and consumption, exploring how changes in agriculture and society have contributed to food waste. Economic lessons can focus on the costs of food waste and its impact on global food security. Cultural studies can delve into different food practices around the world and how they influence waste, promoting a global perspective on the issue.

Language Arts:

Language arts classes can incorporate food waste topics through reading, writing, and discussions. Students can read articles, books, and case studies about food waste, engaging in critical analysis and reflection. Writing assignments can include persuasive essays, research papers, and creative projects that encourage students to articulate their thoughts on food waste and propose solutions. Discussions and debates can further develop their communication skills and foster a collaborative learning environment.

By leveraging cross-curricular opportunities, educators can create a rich and varied educational experience that highlights the relevance of food waste across different fields of study. This approach not only enhances students' knowledge and skills but also encourages them to think critically and creatively about addressing the issue.

Lesson Planning and Resources

Effective lesson planning and the use of diverse resources are essential for integrating food waste topics into the curriculum. Teachers can design engaging and informative lessons that capture students' interest and promote active learning.

Creating Lesson Plans:

When creating lesson plans, teachers should align food waste topics with their curriculum standards and learning objectives. Each lesson should have clear goals, such as understanding the causes of food waste, analyzing its impacts, and exploring solutions. Activities should be varied and interactive, catering to different learning styles and encouraging student participation.

For example, a lesson on the environmental impacts of food waste could include a combination of lectures, multimedia presentations, and hands-on activities. Students might watch a documentary on food waste, participate in a classroom discussion, and then engage in a composting project to apply what they've learned. This multifaceted approach ensures that students are not only informed but also actively involved in the learning process.

Utilizing Resources:

There are numerous resources available to support the integration of food waste topics into the curriculum. Teachers can access a wide range of materials, including lesson plans, activities, and multimedia resources from educational websites, government agencies, and non-profit organizations. These resources often provide comprehensive guides and tools that can be easily adapted to different grade levels and subject areas.

Guest Speakers and Field Trips:

Inviting guest speakers and organizing field trips can enhance students' understanding of food waste. Guest speakers, such as environmental scientists, farmers, and representatives from food recovery organizations, can provide expert insights and firsthand experiences. Field trips to farms, food processing facilities, and waste management centers can offer practical, real-world learning opportunities that reinforce classroom lessons.

Collaborative Projects:

Collaborative projects are another effective way to integrate food waste topics. Group activities encourage teamwork, communication, and problem-solving skills. For instance, students can work together to conduct a food waste audit of their school, develop a campaign to reduce waste, or create a community garden that utilizes composted food scraps. These projects not only deepen their understanding of food waste but also foster a sense of responsibility and community engagement.

Assessment and Reflection:

Assessment and reflection are important components of lesson planning. Teachers should incorporate various assessment methods, such as quizzes, essays, presentations, and projects, to evaluate students' understanding and progress. Reflection activities, such as journals and discussions, allow students to process what they've learned and consider how they can apply their knowledge to reduce food waste in their own lives.

By thoughtfully planning lessons and utilizing a variety of resources, teachers can effectively integrate food waste education into their curriculum. This approach not only enriches students' learning experiences but also empowers them to become informed and active participants in the fight against food waste.

Age-Appropriate Learning Objectives

Teaching about food waste requires tailoring the content and activities to the age and developmental stage of the students. By focusing on age-appropriate learning objectives, educators can ensure that students at different grade levels engage with the material in a meaningful and relevant way.

Primary School Focus

For primary school students, the emphasis should be on building foundational knowledge about food waste and fostering an early appreciation for sustainability. Young learners are naturally curious and can grasp basic concepts through interactive and engaging activities.

Understanding Basic Concepts:

Primary school students should start with the basics of what food waste is and why it matters. Lessons can introduce the idea that food waste occurs when food that could be eaten is thrown away. Teachers can use simple language and relatable examples, such as unfinished lunches or spoiled fruits at home, to explain these concepts. Visual aids, like pictures and videos, can help illustrate these ideas and make them more accessible.

Interactive Activities:

Hands-on activities are particularly effective for young learners. Composting projects can be an excellent way to teach students about the decomposition process and the benefits of recycling organic waste. Students can collect food scraps from their snacks and lunches, observe the composting process over time, and learn how compost can be used to nourish plants. This activity not only teaches them about food waste but also introduces basic scientific principles.

Storytelling and Role-Playing:

Storytelling and role-playing can make the topic of food waste more engaging for primary students. Teachers can read stories or show videos featuring characters who learn about and tackle food waste. Role-playing activities, where students act out scenarios involving food waste and its prevention, can help them internalize the lessons and understand their role in reducing waste.

Basic Math and Science Integration:

Integrating basic math and science lessons with food waste topics can reinforce students' understanding. For example, students can count and categorize different types of food waste they produce in a week, then create simple charts or graphs to visualize their data. This activity combines math skills with practical knowledge about food waste.

Encouraging Positive Behaviors:

At this age, it is crucial to encourage positive behaviors that reduce food waste. Teachers can implement classroom practices such as "no waste" snack times, where students are encouraged to bring only what they can eat, and to take home any leftovers. Reward systems and recognition can motivate students to adopt and maintain these behaviors.

By focusing on these age-appropriate learning objectives, primary school teachers can lay the groundwork for a lifelong commitment to reducing food waste and promoting sustainability.

Secondary School Focus

For secondary school students, the focus should shift to a deeper understanding of the causes and impacts of food waste, as well as the development of critical thinking and problem-solving skills. Older students are capable of engaging with more complex concepts and can take a more active role in addressing food waste.

In-Depth Exploration of Causes and Impacts:

Secondary school students should explore the various causes of food waste in greater detail. Lessons can cover agricultural practices, supply chain inefficiencies, and consumer behavior, providing students with a comprehensive understanding of where and why food waste occurs. Discussions on the environmental, economic, and social impacts of food waste can help students grasp the broader consequences of the issue.

Research and Analysis:

Engaging students in research projects can deepen their understanding and analytical skills. For instance, students can conduct research on the food waste practices in their community, gather data through surveys or interviews, and analyze their findings. This process teaches them to critically evaluate information and draw evidence-based conclusions.

Problem-Solving and Innovation:

Secondary students should be encouraged to develop and propose solutions to food waste. This can involve project-based learning where students design and implement initiatives to reduce waste in their school or community. Examples include organizing food drives, creating awareness campaigns, or developing apps that help people manage their food purchases and consumption more efficiently.

Cross-Disciplinary Connections:

At this level, integrating food waste education across different subjects can provide a richer learning experience. In science classes, students can study the biological processes involved in composting and the environmental benefits. In economics, they can analyze the cost implications of food waste and explore economic models that promote sustainability. Language arts classes can focus on persuasive writing, where students create compelling arguments for reducing food waste.

Critical Discussions and Debates:

Facilitating critical discussions and debates on food waste topics can enhance students' understanding and engagement. Teachers can present case studies or current events related to food waste and encourage students to debate different perspectives and solutions. This helps develop their critical thinking and public speaking skills.

Service Learning and Community Engagement:

Service learning projects that involve students in real-world efforts to combat food waste can be highly impactful. Partnering with local organizations, students can participate in food recovery programs, volunteer at food banks, or help set up community composting initiatives. These experiences not only reinforce classroom learning but also instill a sense of civic responsibility and community involvement.

By addressing these age-appropriate learning objectives, secondary school educators can equip students with the knowledge, skills, and motivation to actively contribute to the reduction of food waste and the promotion of sustainability.

Developing Critical Thinking and Problem-Solving Skills

Critical thinking and problem-solving skills are essential for students to effectively address complex issues like food waste. By fostering these skills, educators can empower students to analyze problems, develop innovative solutions, and take meaningful action.

Activities and Projects

Engaging students in activities and projects is a powerful way to develop their critical thinking and problem-solving skills. These hands-on experiences encourage students to apply what they have learned, think creatively, and work collaboratively to find solutions.

Food Waste Audits:

One effective activity is conducting a food waste audit. Students can assess the amount and types of food waste generated in their school cafeteria or at home. This involves collecting data on the volume of food wasted, categorizing the types of waste, and identifying the reasons for disposal. Students can then analyze the data to understand patterns and pinpoint areas for improvement. This activity not only develops analytical skills but also raises awareness about the scale of food waste and its causes.

Research Projects:

Research projects allow students to delve deeper into specific aspects of food waste. For example, students can investigate the environmental impacts of food waste, the economic costs, or the social implications. They can use various sources, such as academic articles, reports, and interviews, to gather information. Students can then present their findings in written reports, presentations, or posters. This process enhances their research, critical thinking, and communication skills.

Problem-Solving Challenges:

Problem-solving challenges can stimulate students' creativity and critical thinking. Teachers can present real-world scenarios related to food waste and ask students to devise solutions. For instance, students might be tasked with creating a plan to reduce food waste in their school cafeteria. They would need to consider various factors, such as menu planning, portion sizes, and waste management practices. By working through these challenges, students learn to approach problems systematically and develop practical solutions.

Debates and Discussions:

Organizing debates and discussions on food waste topics can sharpen students' critical thinking and public speaking skills. Teachers can present controversial issues, such as the effectiveness of different food waste reduction strategies, and have students debate the pros and cons. Through these activities, students learn to construct logical arguments, consider multiple perspectives, and articulate their ideas clearly.

Collaborative Projects:

Collaborative projects encourage teamwork and problem-solving. Students can work in groups to design and implement initiatives to reduce food waste. For example, they could create a composting program, organize a food donation drive, or develop educational materials to raise awareness. These projects require students to plan, coordinate, and execute their ideas, fostering a range of skills from project management to critical thinking.

Encouraging Student-Led Initiatives

Student-led initiatives are an effective way to foster a sense of ownership and responsibility in students. By encouraging students to take the lead on projects and campaigns, educators can cultivate their leadership, problem-solving, and critical thinking skills.

Identifying Issues:

Encouraging students to identify food waste issues in their school or community is a good starting point for student-led initiatives.

Teachers can facilitate brainstorming sessions where students discuss and prioritize the problems they see. This process helps students develop their analytical skills as they assess the significance and impact of different issues.

Planning and Implementation:

Once students have identified an issue, they can plan and implement a project to address it. This involves setting goals, outlining steps, and assigning tasks. For example, students might decide to start a campaign to reduce food waste in the school cafeteria. They would need to develop a plan that includes creating awareness posters, organizing events, and monitoring progress. This process teaches students to think strategically, manage resources, and solve problems collaboratively.

Monitoring and Evaluation:

Monitoring and evaluating the success of their initiatives is a crucial part of the learning process. Students should establish criteria for success and regularly assess their progress. For instance, if their goal is to reduce food waste in the cafeteria, they could track the amount of waste before and after implementing their project. This helps students develop critical thinking skills as they analyze data, reflect on their methods, and make adjustments to improve outcomes.

Reflective Practices:

Encouraging students to engage in reflective practices can deepen their understanding and enhance their problem-solving skills. After completing a project, students can reflect on what worked well, what challenges they faced, and what they could do differently in the future. Reflective journals, group discussions, and presentations can provide opportunities for students to share their experiences and insights. This reflection helps them to internalize lessons learned and apply them to future projects.

Leadership Development:

Student-led initiatives provide valuable opportunities for leadership development. Students take on roles such as project managers, team leaders, and coordinators, allowing them to practice and hone their leadership skills. These experiences build confidence and empower students to take initiative in addressing food waste and other issues.

Community Engagement:

Encouraging students to engage with the community can extend the impact of their initiatives. Students can partner with local organizations, businesses, and government agencies to implement their projects. For example, they might collaborate with a local food bank to organize a food donation drive or work with a waste management company to set up a composting program. These partnerships provide real-world experience and help students understand the broader context of food waste.

By fostering student-led initiatives, educators can create a dynamic and empowering learning environment that promotes critical thinking, problem-solving, and leadership. These experiences prepare students to tackle food waste and other challenges with confidence and creativity.

Chapter 2: Practical Classroom Activities

Practical classroom activities are essential for bringing the topic of food waste to life and engaging students in meaningful learning experiences. This chapter provides a range of hands-on, interactive activities designed to help students understand the causes and impacts of food waste, as well as develop skills to reduce it. By incorporating these activities into the curriculum, educators can foster a deeper understanding of food waste while promoting critical thinking, problem-solving, and collaboration. From composting projects to digital tools and collaborative initiatives, these activities offer diverse and effective ways to engage students and inspire action.

Hands-On Learning

Hands-on learning activities are a powerful way to engage students in the topic of food waste. These activities not only make learning more interactive and enjoyable but also help students develop practical skills and a deeper understanding of the subject. Two effective hands-on learning activities are composting projects and growing food in the classroom.

Composting Projects

Composting projects are an excellent way to teach students about the decomposition process and the benefits of recycling organic waste. By participating in composting, students can see firsthand how food scraps and other organic materials break down over time and transform into nutrient-rich soil.

Introduction to Composting:

Begin by introducing students to the basics of composting. Explain what composting is, how it works, and why it is important for reducing food waste and enriching soil. Discuss the types of materials that can be composted, such as fruit and vegetable scraps, coffee grounds, eggshells, and yard waste. Emphasize the environmental benefits of composting, including reducing landfill waste and lowering greenhouse gas emissions.

Setting Up a Compost Bin:

Next, guide students in setting up a compost bin. This can be done indoors with a small container or outdoors with a larger bin. Explain the importance of having the right balance of "green" materials (e.g., fruit and vegetable scraps) and "brown" materials (e.g., dry leaves, paper). Show students how to layer these materials and maintain the compost pile by turning it regularly to aerate it and speed up the decomposition process.

Monitoring the Compost:

Students should monitor the compost regularly to observe the changes taking place. Encourage them to take notes and record their observations, such as the temperature of the compost, the presence of different types of organisms (e.g., worms, insects), and the breakdown of materials over time. This hands-on experience helps students understand the science behind decomposition and the factors that influence it.

Using Finished Compost:

Once the compost is ready, students can use it to enrich the soil in a garden or potted plants. This final step completes the cycle of food waste reduction and demonstrates the practical benefits of composting. Encourage students to reflect on the process and the positive impact they have made on the environment.

Connecting to Curriculum:

Composting projects can be integrated into various subjects, such as science (studying decomposition and soil health), math (measuring and recording data), and environmental studies (discussing sustainability and waste reduction). This cross-curricular approach reinforces the learning objectives and makes the activity more meaningful.

By participating in composting projects, students gain a deeper understanding of the decomposition process, learn practical skills for managing organic waste, and develop a sense of responsibility for the environment.

Growing Food in the Classroom

Growing food in the classroom is another hands-on activity that can teach students about the food production process, the importance of fresh produce, and the impact of food waste. This activity provides students with a tangible connection to the food they eat and fosters an appreciation for sustainable practices.

Introduction to Growing Food:

Start by introducing students to the basics of growing food. Discuss the different types of plants that can be grown indoors, such as herbs, lettuce, tomatoes, and radishes. Explain the requirements for plant growth, including sunlight, water, and nutrients. This foundational knowledge sets the stage for the hands-on activity.

Planting Seeds:

Guide students through the process of planting seeds. Provide small pots or containers, potting soil, and seeds. Demonstrate how to fill the pots with soil, plant the seeds at the appropriate depth, and water them. Explain the importance of labeling the pots with the type of plant and the date of planting. This activity helps students develop fine motor skills and understand the initial steps of food production.

Caring for Plants:

Students should take responsibility for caring for their plants by watering them regularly, ensuring they receive adequate sunlight, and monitoring their growth. Encourage students to keep a journal where they record their observations, such as the germination of seeds, the development of leaves, and any challenges they encounter (e.g., pests, insufficient light). This ongoing care teaches students about the needs of living organisms and the importance of consistent attention.

Harvesting and Using Produce:

Once the plants have matured, students can harvest the produce. This step provides a sense of accomplishment and a direct connection to the food they have grown. Encourage students to use the harvested produce in simple recipes, such as salads or herb-infused dishes. This practical application reinforces the value of fresh, home-grown food and reduces reliance on store-bought produce that might otherwise go to waste.

Connecting to Curriculum:

Growing food in the classroom can be linked to subjects such as science (studying plant biology and ecology), health (discussing nutrition and healthy eating), and environmental studies (exploring sustainable agriculture and food systems). By integrating this activity into the curriculum, teachers can provide a comprehensive learning experience that highlights the connections between food production, consumption, and waste.

By engaging in the process of growing food, students gain a deeper appreciation for the effort required to produce food, learn about sustainable practices, and develop practical skills that can be applied at home and in their communities. This hands-on activity not only enhances their understanding of food systems but also fosters a sense of responsibility for reducing food waste and promoting sustainability.

Interactive Lessons

Interactive lessons are an effective way to engage students and make learning about food waste both fun and educational. By incorporating role-playing, simulations, and field trips, teachers can provide students with hands-on experiences that deepen their understanding and inspire action.

Role-Playing and Simulations

Role-playing and simulations offer dynamic and immersive ways for students to explore the complexities of food waste. These activities can help students understand different perspectives, develop problem-solving skills, and apply their knowledge in practical scenarios.

Introduction to Role-Playing:

Begin by explaining the concept of role-playing to students. Describe how they will take on different roles and act out scenarios related to food waste. This can include roles such as farmers, grocery store managers, consumers, and policymakers. Role-playing helps students understand the diverse factors that contribute to food waste and the perspectives of various stakeholders.

Setting the Scene:

Create realistic scenarios for students to role-play. For example, a scenario might involve a grocery store manager trying to reduce food waste while maintaining profitability. Another scenario could involve a family planning meals to minimize food waste and make the most of their groceries. Provide students with background information, objectives, and challenges for each role to guide their actions and decisions.

Conducting the Role-Play:

Divide students into groups and assign roles. Allow them time to prepare and discuss their strategies. As they act out the scenarios, encourage them to consider the impact of their decisions on food waste and other factors, such as cost, convenience, and sustainability. This interactive approach helps students think critically about the trade-offs and complexities involved in reducing food waste.

Debriefing and Reflection:

After the role-play, facilitate a debriefing session where students can reflect on their experiences. Ask them to discuss what they learned, the challenges they faced, and the insights they gained. Encourage them to consider how their actions and decisions could be applied in real life. This reflection helps reinforce the lessons learned and deepens their understanding of food waste management.

Simulations:

Simulations can complement role-playing by providing structured, game-like environments where students can experiment with different strategies to reduce food waste. For example, a computer simulation might allow students to manage a virtual farm or grocery store, making decisions about production, inventory, and waste management. These simulations provide immediate feedback and help students understand the consequences of their actions.

Connecting to Curriculum:

Role-playing and simulations can be integrated into various subjects, such as social studies (exploring the roles of different stakeholders), economics (analyzing the cost implications of food waste), and environmental science (understanding the environmental impacts of waste). By linking these activities to the curriculum, teachers can provide a comprehensive learning experience that highlights the importance of reducing food waste.

Role-playing and simulations are powerful tools for engaging students in interactive and experiential learning. These activities help students develop critical thinking, problem-solving, and collaboration skills while deepening their understanding of food waste and its impacts.

Field Trips to Farms and Waste Facilities

Field trips to farms and waste facilities offer students a firsthand look at the food production and waste management processes. These experiences provide valuable insights into the sources and impacts of food waste and the efforts to mitigate it.

Planning the Field Trip:

Begin by planning a field trip to a local farm, food processing facility, or waste management center. Coordinate with the facility to arrange a tour and ensure that the visit aligns with your educational objectives. Prepare students by discussing the purpose of the trip and what they can expect to see and learn.

Visit to a Farm:

A trip to a farm can help students understand the early stages of the food supply chain and the factors that contribute to food waste at the production level. During the visit, students can observe farming practices, learn about crop management, and see how farmers handle surplus or imperfect produce. They can also learn about the challenges farmers face, such as pests, weather conditions, and market demands, and how these factors influence food waste.

Visit to a Food Processing Facility:

At a food processing facility, students can see how raw agricultural products are transformed into packaged foods. They can learn about the processes involved in sorting, cleaning, packaging, and distributing food. The visit can highlight the points at which food waste occurs during processing, such as due to equipment malfunctions, quality control standards, or spoilage. Students can also learn about efforts to minimize waste, such as recycling by-products and implementing efficient production practices.

Visit to a Waste Management Center:

A trip to a waste management center can provide insights into how food waste is handled after it is discarded. Students can see the different methods of waste disposal, such as landfilling, incineration, and composting. They can learn about the environmental impacts of these methods, such as methane emissions from landfills and the benefits of composting organic waste. The visit can also introduce students to innovative waste reduction initiatives, such as food recovery programs and waste-to-energy technologies.

Post-Trip Reflection and Activities:

After the field trip, facilitate a reflection session where students can share their observations and insights. Encourage them to discuss what they learned about the sources and impacts of food waste and the efforts to reduce it. Students can also engage in follow-up activities, such as writing reports, creating presentations, or developing action plans to reduce food waste in their own lives or communities.

Connecting to Curriculum:

Field trips can be integrated into various subjects, such as science (studying agricultural practices and waste management), geography (exploring the locations and environmental conditions of farms and facilities), and social studies (understanding the roles of different stakeholders in the food supply chain). By linking these experiences to the curriculum, teachers can provide a comprehensive and contextualized learning experience.

Field trips to farms and waste facilities offer students valuable, real-world insights into the food production and waste management processes. These experiences deepen their understanding of the complexities of food waste and inspire them to take action to reduce waste and promote sustainability.

Digital Resources and Tools

Digital resources and tools offer innovative and engaging ways to teach students about food waste. By incorporating online games, apps, virtual tours, and videos into the curriculum, educators can enhance learning experiences and provide students with interactive methods to explore the topic of food waste.

Online Games and Apps

Online games and apps are effective tools for making learning about food waste fun and interactive. These digital resources can help students understand the complexities of food waste, develop problem-solving skills, and motivate them to take action.

Introduction to Online Games and Apps:

Begin by introducing students to various online games and apps designed to educate about food waste. These digital tools can cover a wide range of topics, from managing food production and consumption to understanding the environmental impacts of waste. By engaging with these resources, students can learn through play, which enhances retention and makes learning more enjoyable.

Food Waste Management Games:

Games that simulate food waste management scenarios can provide students with practical insights into the challenges and strategies involved in reducing waste. For example, a game might allow students to run a virtual farm or restaurant, where they must make decisions about purchasing, inventory management, and waste reduction. Through gameplay, students can experiment with different approaches and see the immediate consequences of their decisions, helping them understand the complexities of food waste management.

Educational Apps:

There are numerous educational apps available that focus on food waste. These apps often include interactive features, such as quizzes, challenges, and tracking tools, that engage students and encourage active learning. For instance, an app might allow students to track their food waste at home, set goals for reduction, and receive tips and feedback on how to improve. This hands-on approach helps students apply what they learn in real-life situations, reinforcing positive behaviors and habits.

Interactive Simulations:

Interactive simulations provide a more immersive learning experience by allowing students to explore food waste scenarios in a controlled environment. For example, a simulation might place students in the role of a grocery store manager who must balance inventory, sales, and waste reduction. By making decisions and observing the outcomes, students develop critical thinking and problem-solving skills. These simulations can also include real-world data and scenarios, making the learning experience more relevant and impactful.

Gamification of Learning:

Gamification involves incorporating game-like elements into educational activities to increase engagement and motivation. Teachers can use gamification techniques, such as point systems, badges, and leaderboards, to encourage students to participate in food waste reduction activities. For example, students might earn points for completing tasks like composting, reducing food waste at home, or participating in food recovery programs. This competitive element can motivate students to take action and reinforce their commitment to reducing food waste.

Connecting to Curriculum:

Online games and apps can be integrated into various subjects, such as science (exploring the environmental impacts of food waste), economics (understanding the financial costs of waste), and health (promoting healthy eating habits and reducing food waste). By linking these digital resources to the curriculum, teachers can provide a comprehensive learning experience that enhances students' understanding of food waste and encourages them to take action.

Virtual Tours and Videos

Virtual tours and videos offer students the opportunity to explore food waste topics in a visually engaging and accessible way. These digital resources can provide insights into food production, waste management, and sustainability practices from the comfort of the classroom.

Introduction to Virtual Tours and Videos:

Begin by introducing students to the concept of virtual tours and educational videos. Explain how these resources can provide a closer look at food waste processes, from farm to table, and beyond. Virtual tours can offer immersive experiences that are not possible within the confines of the classroom, while videos can present complex information in a clear and engaging manner.

Virtual Tours of Farms and Food Processing Facilities:

Virtual tours of farms and food processing facilities can give students an inside look at the food supply chain. These tours can highlight key stages where food waste occurs, such as harvesting, processing, and packaging. Students can observe how food is grown, harvested, and processed, and learn about the challenges and best practices for minimizing waste. For example, a virtual tour of an organic farm might show sustainable farming practices, while a tour of a food processing plant could demonstrate how technology is used to reduce waste.

Waste Management Facility Tours:

Virtual tours of waste management facilities can provide insights into how food waste is handled after it is discarded. Students can see the different methods of waste disposal, such as composting, anaerobic digestion, and landfill management. These tours can also highlight innovative waste reduction initiatives, such as food recovery programs and waste-to-energy projects. By understanding the end-of-life stages of food waste, students can appreciate the importance of waste reduction and sustainable waste management practices.

Educational Videos:

Educational videos are a versatile tool for teaching about food waste. These videos can cover a wide range of topics, from the environmental impacts of food waste to practical tips for reducing waste at home. Teachers can use videos to introduce new concepts, reinforce lessons, and provide visual examples that enhance understanding. For example, a video on composting can demonstrate the process step-by-step, while a documentary on food waste can provide a broader perspective on the issue.

Interactive Video Platforms:

Interactive video platforms, such as Edpuzzle or Nearpod, allow teachers to create customized video lessons with embedded questions, quizzes, and discussions. These platforms can make video content more engaging and interactive, helping students stay focused and retain information. Teachers can use interactive videos to assess students' understanding, prompt critical thinking, and facilitate classroom discussions on food waste topics.

Connecting to Curriculum:

Virtual tours and videos can be integrated into various subjects, such as geography (exploring the locations and conditions of farms and facilities), science (understanding the biological and technological aspects of food production and waste management), and social studies (examining the roles of different stakeholders in the food supply chain). By linking these digital resources to the curriculum, teachers can provide a rich and contextualized learning experience that enhances students' understanding of food waste.

Virtual tours and videos offer engaging and accessible ways for students to explore food waste topics. These digital resources can provide valuable insights, stimulate curiosity, and inspire action, helping students develop a deeper understanding of food waste and its impacts.

Collaborative Projects

Collaborative projects are an effective way to engage students in learning about food waste while fostering teamwork, communication, and problem-solving skills. By working together on group presentations and community involvement initiatives, students can explore the issue of food waste more deeply and develop practical solutions.

Group Presentations

Group presentations provide students with the opportunity to research, analyze, and share their findings on food waste. This collaborative activity helps students develop their research, organizational, and public speaking skills while promoting a deeper understanding of the topic.

Introduction to Group Presentations:

Start by explaining the purpose and structure of group presentations. Divide the class into small groups and assign each group a specific topic related to food waste, such as the causes of food waste, its environmental impacts, or strategies for reducing waste. Provide guidelines on how to research their topic, organize their findings, and create an engaging presentation.

Research and Preparation:

Each group should begin by conducting thorough research on their assigned topic. Encourage students to use a variety of sources, such as academic articles, government reports, and credible websites. They should gather information, take notes, and identify key points to include in their presentation. This research process helps students develop critical thinking and information literacy skills.

Organizing the Presentation:

Once the research is complete, groups should organize their findings into a coherent and logical presentation. They can use visual aids, such as slides, posters, or infographics, to enhance their presentation and make the information more accessible. Each group member should be assigned specific sections to present, ensuring that everyone has an opportunity to contribute and practice their public speaking skills.

Practicing and Refining:

Encourage groups to practice their presentations multiple times to refine their delivery and ensure a smooth flow. They should focus on clear communication, proper pacing, and effective use of visual aids. Practicing also helps build confidence and reduce anxiety about speaking in front of an audience.

Delivering the Presentation:

On the day of the presentations, each group should present their findings to the class. Encourage the audience to ask questions and provide feedback, fostering an interactive and supportive learning environment. This experience helps students develop their public speaking skills, build confidence, and learn to communicate complex information effectively.

Reflecting on the Experience:

After the presentations, facilitate a reflection session where students can discuss what they learned from the activity. Encourage them to reflect on the research process, the challenges they faced, and the skills they developed. This reflection helps reinforce the lessons learned and highlights the importance of collaboration and effective communication.

By engaging in group presentations, students can deepen their understanding of food waste, develop valuable skills, and learn the importance of working together to address complex issues.

Community Involvement

Community involvement projects provide students with the opportunity to apply their knowledge about food waste in real-world contexts and make a positive impact in their communities. These projects help students develop a sense of civic responsibility and understand the importance of community engagement in addressing food waste.

Introduction to Community Involvement:

Begin by explaining the concept of community involvement and its significance in addressing food waste. Discuss different types of projects students can undertake, such as organizing food drives, setting up composting programs, or partnering with local organizations to reduce food waste. Encourage students to brainstorm ideas and choose a project that aligns with their interests and the needs of their community.

Planning the Project:

Once a project is chosen, students should develop a detailed plan outlining the goals, steps, and timeline. They should identify the resources needed, assign tasks, and establish a schedule for completing each phase of the project. This planning process helps students develop organizational and project management skills.

Implementing the Project:

As students begin to implement their project, they should work collaboratively to carry out the tasks and overcome any challenges that arise. For example, if they are organizing a food drive, they will need to coordinate with local businesses, promote the event, and collect and distribute the donations. Throughout the implementation process, students learn to work as a team, communicate effectively, and adapt to changing circumstances.

Engaging the Community:

Community involvement projects often require students to engage with various stakeholders, such as local residents, businesses, and organizations. Encourage students to reach out to these stakeholders, explain their project, and seek support and collaboration. This engagement helps students build interpersonal skills and understand the value of community partnerships in addressing food waste.

Monitoring and Evaluating:

Students should regularly monitor the progress of their project and evaluate its impact. This can involve tracking metrics, such as the amount of food collected in a drive or the participation rate in a composting program. Encourage students to reflect on their successes and challenges and consider ways to improve the project in the future. This evaluation process helps students develop critical thinking and problem-solving skills.

Sharing Results and Celebrating Success:

Once the project is complete, students should share the results with their community. This can be done through presentations, reports, or social media updates. Celebrating the success of the project helps build a sense of accomplishment and reinforces the importance of community involvement in reducing food waste. It also provides an opportunity to thank stakeholders and encourage continued efforts to address food waste.

Reflecting on the Experience:

Facilitate a reflection session where students can discuss their experiences and what they learned from the community involvement project. Encourage them to reflect on the skills they developed, the challenges they faced, and the impact they made. This reflection helps students internalize the lessons learned and appreciate the value of community engagement in addressing food waste.

By participating in community involvement projects, students can apply their knowledge in real-world contexts, develop practical skills, and make a positive impact in their communities. These projects help students understand the importance of civic responsibility and the role of community collaboration in addressing food waste.

Chapter 3: Reducing Food Waste in Schools

Reducing food waste in schools is a crucial step toward fostering a culture of sustainability and environmental responsibility among students. Schools are not only places of learning but also significant contributors to food waste, from cafeterias to classroom activities. This chapter explores various strategies and initiatives that can be implemented to minimize food waste in the school environment. By engaging students, staff, and the community in these efforts, schools can lead by example and make a substantial impact on reducing food waste. Through school-wide initiatives, student-led campaigns, and parent and community engagement, this chapter provides practical solutions and inspiration for schools committed to reducing their food waste footprint.

School-Wide Initiatives

Implementing school-wide initiatives to reduce food waste can significantly impact the overall waste generated in educational institutions. By engaging the entire school community in these efforts, schools can foster a culture of sustainability and environmental stewardship. Two effective strategies for reducing food waste are cafeteria waste audits and food sharing programs.

Cafeteria Waste Audits

Cafeteria waste audits are a valuable tool for understanding the amount and types of food waste generated in school cafeterias. By systematically assessing food waste, schools can identify problem areas and develop targeted strategies to reduce waste.

Introduction to Cafeteria Waste Audits:

Begin by explaining the purpose and importance of conducting cafeteria waste audits. Highlight how these audits can provide insights into food consumption patterns, identify sources of waste, and inform strategies for waste reduction. Emphasize that involving students in the audit process can enhance their understanding of food waste issues and encourage them to take ownership of the solutions.

Planning the Audit:

To conduct a cafeteria waste audit, start by assembling a team that includes students, teachers, and cafeteria staff. Assign specific roles and responsibilities, such as data collection, analysis, and reporting. Develop a detailed plan outlining the objectives, methodology, and timeline for the audit. Determine the duration of the audit period, which can range from a single day to a week, depending on the school's schedule and resources.

Collecting Data:

During the audit, collect data on the types and quantities of food waste generated in the cafeteria. Set up waste sorting stations where students can separate food waste into categories, such as uneaten food, plate waste, and kitchen scraps. Weigh each category of waste and record the data. Encourage students to observe and note common reasons for food waste, such as oversized portions, unappealing meals, or lack of awareness about waste reduction practices.

Analyzing Results:

After the data collection period, analyze the results to identify patterns and trends. Calculate the total amount of food waste generated and break it down by category. Identify the most common sources of waste and consider potential underlying causes. For example, if a significant portion of waste consists of uneaten vegetables, it may indicate that students find them unappetizing or that portion sizes are too large.

Developing Action Plans:

Based on the audit findings, develop targeted action plans to reduce food waste in the cafeteria. These plans may include adjusting portion sizes, improving menu planning to offer more appealing and nutritious options, and implementing educational campaigns to raise awareness about the importance of reducing food waste. Engage students in brainstorming and implementing these solutions, fostering a sense of responsibility and involvement.

Monitoring and Adjusting:

Regularly monitor the impact of the implemented strategies and make adjustments as needed. Conduct follow-up audits to assess the effectiveness of the action plans and identify any new areas for improvement. Continuously engage the school community in the efforts to reduce food waste and celebrate successes to maintain momentum and enthusiasm.

By conducting cafeteria waste audits, schools can gain valuable insights into their food waste patterns and develop targeted strategies to reduce waste. This initiative not only helps minimize environmental impact but also fosters a culture of sustainability within the school community.

Food Sharing Programs

Food sharing programs are another effective strategy for reducing food waste in schools. These programs facilitate the redistribution of surplus food to those in need, ensuring that edible food does not go to waste.

Introduction to Food Sharing Programs:

Introduce the concept of food sharing programs and explain their benefits. Highlight how these programs can help reduce food waste, support food-insecure individuals, and promote a sense of community and social responsibility. Emphasize that food sharing programs can be a practical and impactful way for schools to address food waste.

Establishing a Food Sharing Program:

To establish a food sharing program, start by identifying potential sources of surplus food within the school, such as uneaten cafeteria meals, unopened packaged items, and excess food from events and activities. Collaborate with cafeteria staff, teachers, and students to collect and organize surplus food. Ensure that all food items meet safety and quality standards for redistribution.

Partnering with Local Organizations:

Form partnerships with local food banks, shelters, and community organizations to distribute the surplus food. These organizations can provide guidance on food safety protocols, storage, and transportation. Establish a regular schedule for food pickups or deliveries and maintain clear communication with the partner organizations to ensure smooth operations.

Promoting the Program:

Raise awareness about the food sharing program within the school community. Use posters, announcements, and social media to inform students, staff, and parents about the initiative and its goals. Encourage students to participate by donating unopened and uneaten food items from their lunches. Highlight the positive impact of the program on reducing food waste and supporting those in need.

Engaging Students:

Involve students in the food sharing program by creating volunteer opportunities. Students can help collect, sort, and package surplus food, as well as assist with logistics and communication efforts. This hands-on involvement fosters a sense of ownership and responsibility and provides valuable lessons in empathy and community service.

Monitoring and Evaluating:

Regularly monitor the progress and impact of the food sharing program. Track the amount of food collected and distributed, and gather feedback from partner organizations and recipients. Use this information to evaluate the program's effectiveness and make any necessary adjustments. Celebrate milestones and successes to maintain enthusiasm and commitment within the school community.

By implementing food sharing programs, schools can significantly reduce food waste while supporting food-insecure individuals and fostering a sense of community and social responsibility. This initiative not only addresses the immediate issue of food waste but also promotes long-term sustainability and empathy among students.

School-wide initiatives such as cafeteria waste audits and food sharing programs are powerful strategies for reducing food waste in schools. These initiatives engage the entire school community, foster a culture of sustainability, and provide students with valuable learning experiences that extend beyond the classroom.

Student-Led Campaigns

Student-led campaigns are a dynamic way to engage students in tackling food waste. By taking ownership of these initiatives, students can develop leadership skills, raise awareness, and drive meaningful change within their school and community. Two impactful types of student-led campaigns are awareness campaigns and fundraising and advocacy efforts.

Awareness Campaigns

Awareness campaigns are designed to educate and inform the school community about the issue of food waste. By raising awareness, students can inspire their peers, teachers, and families to take action and reduce food waste.

Introduction to Awareness Campaigns:

Begin by explaining the importance of raising awareness about food waste. Highlight how increasing knowledge and understanding of the issue can lead to behavior change and a reduction in food waste. Emphasize that students are in a unique position to influence their peers and the broader community.

Planning the Campaign:

Start by forming a student committee to lead the campaign. This committee should include students from different grade levels and backgrounds to ensure diverse perspectives and ideas. The committee should begin by researching the issue of food waste, gathering facts, statistics, and real-life examples to support their message.

Next, the committee should develop clear objectives for the campaign, such as increasing awareness about the environmental impact of food waste, promoting waste reduction practices, or encouraging participation in school-wide initiatives like composting or food sharing programs.

Creating Engaging Materials:

Students should create a variety of materials to convey their message. These can include posters, flyers, social media posts, videos, and presentations. Visual aids like infographics can be particularly effective in presenting statistics and facts in an engaging way. Encourage students to use creative and attention-grabbing designs to capture their audience's interest.

Implementing the Campaign:

The campaign should be implemented over a set period, such as a week or a month, to maintain momentum and visibility. Students can organize events like assemblies, workshops, and classroom presentations to share their message. They can also set up information booths in high-traffic areas of the school, such as the cafeteria or main entrance, where they can interact with their peers and distribute materials.

Engaging the School Community:

Involve the entire school community in the campaign by encouraging participation and feedback. Students can organize competitions, such as poster contests or food waste reduction challenges, to engage their peers actively. Teachers and staff can support the campaign by incorporating food waste topics into their lessons and discussions.

Measuring Impact:

To gauge the success of the campaign, students should collect data on its impact. This can include surveys to assess changes in awareness and attitudes, tracking participation in campaign activities, and measuring any reduction in food waste during the campaign period. Analyzing this data can provide valuable insights into the effectiveness of the campaign and identify areas for improvement.

By leading awareness campaigns, students can significantly impact reducing food waste within their school and community. These campaigns not only raise awareness but also empower students to take initiative and become advocates for sustainability.

Fundraising and Advocacy

Fundraising and advocacy campaigns are another powerful way for students to address food waste. These efforts can support food waste reduction initiatives, provide resources for those in need, and influence policy changes.

Introduction to Fundraising and Advocacy:

Explain the dual goals of fundraising and advocacy campaigns: to raise funds for food waste reduction initiatives and to advocate for changes in policies and practices that contribute to food waste. Highlight the importance of combining these efforts to create a comprehensive approach to addressing the issue.

Planning the Campaign:

Form a student committee to lead the fundraising and advocacy campaign. This committee should begin by identifying specific goals, such as raising funds for a local food bank, supporting a school composting program, or advocating for changes in school policies related to food waste. They should also research potential funding sources, such as grants, donations, or partnerships with local businesses.

Developing a Fundraising Strategy:

The committee should develop a clear and detailed fundraising strategy. This can include organizing events like bake sales, charity runs, or benefit concerts. Online fundraising platforms, such as GoFundMe or Kickstarter, can also be used to reach a broader audience. Encourage students to use creative approaches to attract donors and highlight the importance of their cause.

Engaging the Community:

To maximize the impact of the fundraising efforts, students should actively engage the school and local community. This can involve collaborating with other student organizations, reaching out to local businesses for sponsorships, and promoting the campaign through social media and local media outlets. Engaging the community not only helps raise funds but also builds a network of support for ongoing food waste reduction efforts.

Advocacy Efforts:

In addition to fundraising, the campaign should include advocacy efforts to promote policy changes and practices that reduce food waste. Students can organize petitions, write letters to school administrators or local government officials, and participate in public meetings or hearings. Advocacy can also involve educating peers and the community about food waste policies and encouraging them to take action.

Implementing and Monitoring:

As funds are raised, the committee should ensure they are used effectively to support the identified goals. This might involve purchasing composting bins, funding educational materials, or supporting food recovery programs. Regularly monitor the progress of both the fundraising and advocacy efforts to ensure they are on track and achieving their objectives.

Evaluating Success:

After the campaign, students should evaluate its success by measuring the funds raised, the impact of advocacy efforts, and the overall engagement of the school and community. This evaluation can provide insights into what worked well and what could be improved in future campaigns. Sharing the results with the school community can also help maintain momentum and support for ongoing food waste reduction initiatives.

By leading fundraising and advocacy campaigns, students can play a crucial role in reducing food waste and supporting sustainability efforts. These campaigns not only provide practical resources and influence policy changes but also empower students to become active and engaged citizens.

Engaging Parents and the Community

Engaging parents and the broader community is essential for the success of food waste reduction initiatives in schools. By involving these key stakeholders, schools can extend the impact of their efforts beyond the classroom and foster a community-wide commitment to sustainability. This section explores how workshops and events, along with effective communication strategies, can be used to engage parents and the community in food waste reduction efforts.

Workshops and Events

Workshops and events are effective ways to involve parents and community members in food waste reduction initiatives. These activities provide opportunities for education, collaboration, and hands-on involvement, making them powerful tools for fostering community engagement.

Introduction to Workshops and Events:

Begin by explaining the purpose of organizing workshops and events. Highlight how these activities can raise awareness about food waste, provide practical tips for waste reduction, and create a sense of community around sustainability efforts. Emphasize that involving parents and community members in these initiatives can lead to more significant and lasting impacts.

Planning and Organizing Workshops:

To plan a successful workshop, start by identifying specific topics related to food waste that would be of interest to parents and the community. Potential topics include composting at home, meal planning and portion control, preserving and storing food, and understanding food date labels. Collaborate with local experts, such as nutritionists, chefs, and environmental educators, to provide engaging and informative presentations.

Workshops should be interactive and hands-on, allowing participants to practice new skills and ask questions. For example, a composting workshop might include a demonstration of different composting methods, while a meal planning workshop could involve interactive exercises on creating balanced and waste-free meal plans. Provide participants with resources and take-home materials, such as guides, recipes, and composting kits, to reinforce the workshop content.

Organizing Community Events:

In addition to workshops, community events can be an effective way to engage a broader audience. Consider organizing events such as food waste awareness fairs, community clean-up days, or zero-waste potlucks. These events can feature activities like cooking demonstrations, waste audits, and educational booths, creating a festive and informative atmosphere.

Events should be well-publicized and accessible to all community members. Use multiple channels to promote the event, including school newsletters, social media, local newspapers, and community bulletin boards. Encourage students to invite their families and friends, and consider partnering with local businesses and organizations to increase participation and support.

Involving Students:

Involve students in the planning and execution of workshops and events to foster a sense of ownership and responsibility. Students can help design promotional materials, assist with event logistics, and even lead parts of the workshops or activities. Their involvement not only enriches their learning experience but also demonstrates the school's commitment to student-led initiatives.

Evaluating Success:

After the workshop or event, gather feedback from participants to evaluate its success and identify areas for improvement. Use surveys, comment cards, and informal conversations to collect insights on what participants found most valuable and what could be enhanced in future activities. Share the feedback with the planning team and use it to inform future workshops and events.

By organizing engaging and informative workshops and events, schools can build a strong network of support for food waste reduction initiatives and inspire lasting change within the community.

Communication Strategies

Effective communication strategies are crucial for engaging parents and the community in food waste reduction efforts. By using clear, consistent, and targeted communication, schools can raise awareness, encourage participation, and build a collective commitment to sustainability.

Introduction to Communication Strategies:

Begin by explaining the importance of communication in engaging parents and the community. Highlight how effective communication can inform, motivate, and mobilize individuals to take action against food waste. Emphasize that a well-planned communication strategy can amplify the impact of school initiatives and foster a culture of sustainability.

Developing a Communication Plan:

Start by developing a comprehensive communication plan that outlines the key messages, target audiences, and channels for communication. Identify the main objectives of the communication efforts, such as increasing awareness about food waste, promoting school initiatives, and encouraging community participation.

Key Messages:

Craft clear and compelling key messages that convey the importance of reducing food waste and the benefits of participating in school initiatives. Messages should be tailored to different audiences, such as parents, students, and community members. For example, a message aimed at parents might focus on practical tips for reducing food waste at home, while a message for the broader community might highlight the environmental and social impacts of food waste.

Communication Channels:

Utilize a variety of communication channels to reach different segments of the community. Common channels include school newsletters, social media, email, the school website, local newspapers, and community bulletin boards. Each channel has its strengths and can be used to target specific audiences effectively.

Social Media:

Social media platforms like Facebook, Twitter, and Instagram are powerful tools for reaching a broad audience and engaging them in food waste reduction efforts. Use social media to share updates, promote events, and highlight success stories. Create engaging content, such as infographics, videos, and testimonials, to capture attention and encourage sharing. Encourage students and parents to follow and engage with the school's social media accounts to build a sense of community and support.

Newsletters and Email:

School newsletters and email communications are effective ways to keep parents informed about food waste initiatives and events. Include regular updates on the progress of school projects, upcoming workshops and events, and practical tips for reducing food waste at home. Personalize communications to make them relevant and engaging for recipients.

Community Partnerships:

Build partnerships with local organizations, businesses, and media outlets to amplify communication efforts. Collaborate with local newspapers and radio stations to feature stories about the school's food waste initiatives. Partner with businesses to promote events and distribute educational materials. These partnerships can extend the reach of communication efforts and enhance community support.

Engaging Students:

Involve students in the communication efforts by encouraging them to create content, such as articles, videos, and social media posts, about food waste initiatives. Student-created content can be particularly effective in engaging their peers and families, as it reflects their perspectives and experiences.

Measuring Impact:

Regularly evaluate the effectiveness of communication strategies by tracking engagement metrics, such as social media interactions, email open rates, and attendance at events. Use surveys and feedback forms to gather insights from parents and community members on what communication methods and messages resonate most. Adjust the communication plan based on these insights to continually improve its effectiveness.

By implementing effective communication strategies, schools can engage parents and the community in food waste reduction efforts, build a collective commitment to sustainability, and create lasting change.

Chapter 4: Monitoring and Evaluating Food Waste

Monitoring and evaluating food waste is a crucial step in understanding and addressing the issue effectively. By systematically tracking and analyzing food waste, schools can identify patterns, measure the impact of their waste reduction initiatives, and continuously improve their strategies. This chapter explores the methods and tools for monitoring food waste, setting goals and benchmarks, and evaluating the success of implemented programs. Through regular assessment and adjustment, schools can ensure that their efforts are making a meaningful difference, fostering a culture of accountability and sustainability within the school community.

Setting Goals and Benchmarks

Establishing clear goals and benchmarks is essential for effectively reducing food waste in schools. By setting measurable objectives and tracking progress, schools can ensure that their initiatives are focused, effective, and continuously improving. This section provides guidance on creating actionable goals and systematically monitoring progress toward achieving them.

Creating Measurable Objectives

Creating measurable objectives is the first step in setting effective goals for reducing food waste. These objectives should be specific, attainable, and aligned with the overall mission of promoting sustainability within the school.

Defining Specific Goals:

Begin by defining clear and specific goals that address the key aspects of food waste reduction. Specific goals provide a clear direction and make it easier to measure progress. For example, a goal might be to reduce cafeteria food waste by 30% over the next school year. Specific goals can also focus on different areas, such as reducing plate waste, increasing composting participation, or improving food donation programs.

Ensuring Goals are Attainable:

While it is important to be ambitious, goals should also be realistic and attainable. Consider the current baseline of food waste in the school and set goals that are challenging yet achievable. Involving stakeholders, such as students, teachers, and cafeteria staff, in the goal-setting process can provide valuable insights and ensure that the goals are practical and supported by those who will be implementing the changes.

Aligning with Broader Initiatives:

Goals should align with broader sustainability initiatives and the school's mission. This alignment ensures that food waste reduction efforts are integrated into the overall strategy and supported at all levels. For instance, if the school is committed to achieving zero waste, food waste reduction goals should be a critical component of this broader objective.

Creating Actionable Steps:

Once goals are set, break them down into actionable steps. These steps should outline the specific actions needed to achieve the goals. For example, if the goal is to reduce cafeteria food waste, actionable steps might include conducting waste audits, adjusting portion sizes, implementing a food sharing program, and educating students about waste reduction. Clearly defined steps provide a roadmap for achieving the objectives and make it easier to monitor progress.

Setting Timelines and Deadlines:

Establishing timelines and deadlines is crucial for maintaining momentum and accountability. Set specific deadlines for each action step and regularly review progress. For example, the goal might be to implement composting bins in the cafeteria within the first three months, conduct a mid-year waste audit, and achieve the 30% reduction target by the end of the school year. Timelines help ensure that initiatives stay on track and that progress is consistently monitored.

Measuring Success:

Define how success will be measured. Success metrics might include the percentage reduction in food waste, the number of students participating in composting, or the volume of food donated. Clearly defined metrics provide a basis for evaluating progress and determining whether goals are being met. Regularly reviewing these metrics helps identify areas of success and opportunities for improvement.

By creating measurable objectives, schools can set a clear direction for their food waste reduction efforts and establish a framework for monitoring and evaluating progress. This structured approach ensures that initiatives are focused, effective, and continuously improving.

Tracking Progress

Tracking progress is essential for ensuring that food waste reduction initiatives are on track and achieving their intended impact. By systematically monitoring key metrics and regularly reviewing progress, schools can identify areas for improvement and make data-driven decisions.

Establishing Baseline Data:

Begin by establishing baseline data on current food waste levels. Conducting a waste audit provides a snapshot of the amount and types of food waste generated in the school. This baseline data serves as a reference point for measuring progress and setting realistic goals. For example, a waste audit might reveal that the school generates an average of 200 pounds of food waste per week. This information is crucial for setting reduction targets and tracking progress over time.

Developing a Monitoring System:

Develop a system for regularly monitoring food waste. This system should include procedures for collecting and recording data on food waste levels, participation in waste reduction programs, and other relevant metrics. Designate specific individuals or teams responsible for data collection and ensure that they have the necessary tools and training. Regular monitoring helps identify trends, track progress, and make timely adjustments to initiatives.

Using Technology and Tools:

Utilize technology and tools to facilitate data collection and analysis. Digital tools, such as spreadsheets, databases, and specialized software, can streamline the process and improve accuracy. For example, schools can use mobile apps to record food waste data, generate reports, and visualize trends. These tools make it easier to analyze data, identify patterns, and share results with stakeholders.

Conducting Regular Audits:

Regular waste audits are a key component of tracking progress. Conduct audits at regular intervals, such as quarterly or semi-annually, to assess the effectiveness of initiatives and measure progress toward goals. During each audit, collect data on the volume and composition of food waste, identify sources of waste, and evaluate changes compared to the baseline data. Regular audits provide ongoing insights and help ensure that efforts are effective and on track.

Reviewing and Analyzing Data:

Regularly review and analyze the collected data to assess progress and identify areas for improvement. Compare the current data with the baseline and previous audits to determine whether goals are being met. Analyze trends and patterns to understand the factors contributing to changes in food waste levels. This analysis helps identify successful strategies and areas that require additional attention or adjustment.

Engaging Stakeholders:

Engage stakeholders, such as students, teachers, and cafeteria staff, in the tracking process. Share the collected data and progress reports with the school community to foster transparency and accountability. Involving stakeholders in the review and analysis process encourages collaboration and provides valuable insights. It also helps build a sense of ownership and commitment to the food waste reduction efforts.

Adjusting Strategies:

Use the insights gained from tracking progress to adjust strategies and initiatives as needed. If the data indicates that certain efforts are not achieving the desired results, consider alternative approaches or additional interventions. For example, if waste audits reveal that a significant amount of food waste comes from uneaten fruits, the school might adjust portion sizes or explore ways to make the fruits more appealing. Continuously refining and improving strategies based on data ensures that efforts are effective and impactful.

Celebrating Successes:

Regularly celebrate successes and milestones to maintain motivation and momentum. Acknowledge and reward the contributions of students, staff, and the broader community. Celebrating achievements not only reinforces positive behaviors but also builds a sense of pride and ownership in the food waste reduction efforts.

By systematically tracking progress, schools can ensure that their food waste reduction initiatives are effective and continuously improving. This data-driven approach enables schools to make informed decisions, identify successful strategies, and achieve their goals of reducing food waste and promoting sustainability.

Data Collection Methods

Effective data collection methods are essential for understanding and addressing food waste in schools. By gathering accurate and comprehensive data, schools can identify problem areas, track progress, and make informed decisions about their food waste reduction strategies. This section outlines the key methods for collecting data, including surveys and audits, and provides guidance on analyzing the collected data to drive meaningful improvements.

Surveys and Audits

Surveys and audits are two primary methods for collecting data on food waste. These tools provide valuable insights into food waste patterns, behaviors, and perceptions within the school community.

Conducting Surveys:

Surveys are a versatile tool for gathering information from a wide range of stakeholders, including students, teachers, cafeteria staff, and parents. Surveys can be used to assess attitudes and behaviors related to food waste, identify areas of concern, and gather suggestions for improvement.

Designing Effective Surveys:

When designing surveys, ensure that the questions are clear, concise, and relevant to the objectives of the food waste reduction initiatives. Use a mix of question types, such as multiple-choice, Likert scale, and open-ended questions, to capture both quantitative and qualitative data. For example, questions might include:

- How often do you throw away uneaten food from your lunch?

- What types of food do you most commonly waste?

- What factors influence your decision to discard food?

- What suggestions do you have for reducing food waste in the school?

Administering Surveys:

Surveys can be administered through various channels, such as online platforms, paper forms, or in-person interviews. Online surveys are particularly convenient for reaching a large audience and can be easily analyzed using digital tools. Ensure that surveys are anonymous to encourage honest and candid responses. Promote the survey widely and provide incentives, such as participation in a raffle or extra credit, to increase response rates.

Conducting Waste Audits:

Waste audits involve systematically collecting and analyzing food waste data to understand the quantity and types of waste generated. Audits provide a detailed and accurate picture of food waste patterns and can inform targeted interventions.

Planning the Audit:

Begin by planning the audit, including defining the scope, timeline, and methodology. Determine the areas to be audited, such as the cafeteria, classrooms, and staff areas. Decide on the duration of the audit, which can range from a single day to a week, depending on the goals and resources available. Assemble a team of volunteers, including students, teachers, and staff, to assist with the audit.

Collecting Data:

During the audit, set up waste sorting stations where food waste can be separated into different categories, such as uneaten whole items, plate waste, and kitchen scraps. Weigh each category and record the data, noting the specific types of food being discarded. Encourage participants to observe and document the reasons for food waste, such as oversized portions, unappealing meals, or time constraints.

Analyzing Audit Results:

After collecting the data, analyze the results to identify patterns and trends. Calculate the total amount of food waste generated and break it down by category. Identify the most common sources of waste and consider potential underlying causes. This analysis provides a baseline for measuring progress and informs the development of targeted interventions.

By conducting surveys and audits, schools can gather comprehensive data on food waste patterns and behaviors. These tools provide valuable insights that inform the development and implementation of effective waste reduction strategies.

Analyzing Waste Data

Analyzing the data collected from surveys and audits is crucial for understanding the underlying causes of food waste and identifying opportunities for improvement. By systematically analyzing the data, schools can make informed decisions and develop targeted interventions to reduce food waste.

Data Compilation and Organization:

Start by compiling and organizing the data collected from surveys and audits. Use digital tools, such as spreadsheets or specialized software, to create a centralized database where all data can be stored and easily accessed. Organize the data into categories, such as types of food waste, sources of waste, and reasons for waste. This organization makes it easier to analyze and interpret the data.

Quantitative Analysis:

Quantitative analysis involves analyzing numerical data to identify patterns and trends. Calculate key metrics, such as the total amount of food waste generated, the percentage of waste by category, and the average waste per student. Use statistical tools to analyze the data and identify significant correlations or differences. For example, you might find that certain types of food are consistently wasted more than others or that food waste is higher on certain days of the week.

Qualitative Analysis:

Qualitative analysis involves interpreting non-numerical data, such as open-ended survey responses and observational notes from audits. Review the qualitative data to identify common themes and insights. For example, you might find that many students cite unappealing meals as a reason for wasting food or that time constraints during lunch periods contribute to higher waste levels. Qualitative analysis provides a deeper understanding of the factors influencing food waste and complements the quantitative findings.

Identifying Patterns and Trends:

Analyze the data to identify patterns and trends that can inform targeted interventions. Look for recurring issues or areas with the highest levels of waste. For example, if the data reveals that a significant portion of waste comes from uneaten fruits and vegetables, consider strategies to make these items more appealing or adjust portion sizes. Identifying patterns and trends helps prioritize areas for improvement and develop focused strategies.

Developing Data-Driven Strategies:

Use the insights gained from data analysis to develop data-driven strategies for reducing food waste. These strategies should address the specific issues identified in the analysis and leverage the strengths of the school community. For example, if surveys indicate that students lack awareness about the impacts of food waste, consider implementing an educational campaign to raise awareness and promote waste reduction practices.

Setting Targets and Benchmarks:

Based on the analysis, set specific targets and benchmarks for reducing food waste. These targets should be realistic and achievable, considering the baseline data and the identified opportunities for improvement. For example, a target might be to reduce cafeteria food waste by 20% over the next semester. Setting clear targets provides a focus for efforts and a basis for measuring progress.

Monitoring and Adjusting:

Regularly monitor progress toward achieving the targets and adjust strategies as needed. Conduct follow-up surveys and audits to assess the impact of the implemented interventions and identify any new issues. Use the data to refine and improve the strategies continuously. Monitoring and adjusting ensure that efforts remain effective and responsive to changing circumstances.

Reporting and Communicating Findings:

Share the findings of the data analysis with the school community to promote transparency and accountability. Use visual aids, such as charts and graphs, to present the data in an accessible and engaging way. Communicate the progress and successes of the food waste reduction initiatives and highlight the contributions of students, staff, and community members. Effective communication fosters a sense of ownership and encourages ongoing support for the initiatives.

By systematically analyzing waste data, schools can gain valuable insights into the causes of food waste and develop targeted, data-driven strategies to reduce it. This approach ensures that food waste reduction efforts are effective, focused, and continuously improving, ultimately contributing to a more sustainable and environmentally responsible school community.

Reporting and Sharing Results

Effectively reporting and sharing the results of food waste reduction initiatives is crucial for maintaining transparency, fostering a sense of community, and sustaining momentum. By communicating findings and celebrating successes, schools can engage and inspire students, staff, and the broader community to continue their efforts in reducing food waste.

Communicating Findings

Communicating the findings of food waste reduction initiatives ensures that all stakeholders are informed about the progress and impact of these efforts. Clear and consistent communication helps build trust, encourages participation, and highlights the importance of the initiatives.

Preparing the Report:

Begin by compiling the data collected from surveys, audits, and other monitoring activities into a comprehensive report. This report should include key metrics, such as the total amount of food waste reduced, the percentage decrease in waste, and any changes in behaviors or practices. Use visual aids like charts, graphs, and infographics to present the data in an accessible and engaging manner.

Sharing the Report:

Distribute the report to all relevant stakeholders, including students, teachers, parents, and community members. Utilize multiple channels to reach a broad audience, such as school newsletters, emails, social media, and the school website. Ensure that the language is clear and jargon-free to make the findings understandable to everyone.

Highlighting Key Insights:

In addition to presenting the data, highlight key insights and lessons learned from the initiatives. Discuss the factors that contributed to the successes and any challenges encountered along the way. This transparency helps build credibility and fosters a collaborative approach to addressing food waste.

Engaging Stakeholders:

Encourage feedback and input from stakeholders by hosting discussions, forums, or Q&A sessions. This engagement allows the school community to ask questions, share their perspectives, and contribute ideas for future initiatives. By involving stakeholders in the conversation, schools can strengthen their commitment to reducing food waste and foster a sense of collective ownership.

Celebrating Successes

Celebrating successes is an essential part of maintaining enthusiasm and motivation for food waste reduction initiatives. Recognizing and rewarding the efforts of students, staff, and the community reinforces positive behaviors and encourages continued participation.

Acknowledging Contributions:

Begin by acknowledging the contributions of all individuals and groups involved in the initiatives. Highlight the roles of students, teachers, cafeteria staff, and community partners in achieving the goals. Public recognition can be given through announcements, certificates, awards, or special events.

Hosting Celebratory Events:

Organize events to celebrate the achievements of the food waste reduction initiatives. These events can include assemblies, award ceremonies, or special activities that highlight the successes and milestones reached. Use these occasions to showcase the impact of the initiatives, share stories of positive change, and thank everyone for their efforts.

Promoting Success Stories:

Share success stories and testimonials from participants who have made significant contributions to the initiatives. These stories can be featured in newsletters, on social media, or in local media outlets. Highlighting personal experiences and achievements helps to humanize the efforts and inspire others to get involved.

Incentivizing Participation:

Consider offering incentives to encourage ongoing participation in food waste reduction efforts. This can include rewards for meeting specific targets, recognition programs, or opportunities for leadership roles in future initiatives. Incentives help maintain engagement and motivate individuals to continue their efforts.

By effectively reporting findings and celebrating successes, schools can sustain momentum, build a supportive community, and inspire continued commitment to reducing food waste. This approach ensures that food waste reduction initiatives remain a dynamic and integral part of the school culture.

Chapter 5: Food Waste and Sustainability

Understanding the relationship between food waste and sustainability is crucial for fostering long-term environmental stewardship. Food waste not only represents a significant loss of resources but also contributes to environmental degradation and climate change. This chapter explores the interconnectedness of food waste with broader sustainability goals, such as reducing greenhouse gas emissions, conserving water and energy, and promoting ethical food systems. By linking food waste reduction efforts to sustainability, schools can inspire students to adopt sustainable practices, understand the global impacts of their actions, and contribute to a more sustainable future. Through comprehensive education and practical initiatives, this chapter aims to equip students with the knowledge and skills needed to make informed and responsible choices regarding food waste and sustainability.

Linking Food Waste to Sustainability Goals

Linking food waste to sustainability goals is essential for understanding the broader implications of waste reduction and fostering a holistic approach to environmental stewardship. By connecting food waste reduction to global sustainability efforts, such as the United Nations Sustainable Development Goals (UN SDGs), students can gain a deeper appreciation of how their actions contribute to a more sustainable world. This section explores the relevance of the UN SDGs and how food waste reduction intersects with broader sustainability issues.

Understanding the UN SDGs

The United Nations Sustainable Development Goals (UN SDGs) provide a universal framework for addressing the world's most pressing challenges, including poverty, inequality, and environmental degradation. Adopted by all UN Member States in 2015, the 17 SDGs are a call to action for countries, organizations, and individuals to work together toward a more sustainable and equitable future by 2030.

Introduction to the UN SDGs:

Begin by introducing students to the UN SDGs and their significance. Explain that the SDGs encompass a wide range of interconnected issues, from clean water and sanitation to responsible consumption and production. Highlight the holistic nature of the goals, emphasizing that achieving one goal often contributes to progress in others. This interconnected approach reflects the complex and interdependent nature of global challenges.

Key SDGs Related to Food Waste:

Focus on the SDGs that are directly related to food waste. SDG 2 (Zero Hunger), SDG 12 (Responsible Consumption and Production), and SDG 13 (Climate Action) are particularly relevant. Explain how reducing food waste can help achieve these goals:

- SDG 2: Zero Hunger: Reducing food waste contributes to food security by ensuring that more food reaches those in need. Explain that wasted food represents a missed opportunity to feed hungry populations, and highlight initiatives that redistribute surplus food to those in need.

- SDG 12: Responsible Consumption and Production: This goal emphasizes the need for sustainable consumption and production patterns. Discuss how minimizing food waste reduces the demand for agricultural production, conserves natural resources, and promotes efficient use of resources throughout the supply chain.

- SDG 13: Climate Action: Food waste contributes to greenhouse gas emissions, which drive climate change. Explain that reducing food waste can significantly lower these emissions by decreasing the amount of organic waste sent to landfills, where it produces methane, a potent greenhouse gas.

Educational Activities:

Incorporate educational activities that help students understand the SDGs and their relevance to food waste. For example, create a project where students research and present on different SDGs, with a focus on how food waste reduction can support each goal. Encourage students to explore case studies and real-world examples of successful initiatives that align with the SDGs.

Connecting Local Actions to Global Goals:

Help students understand that their actions, no matter how small, contribute to global sustainability efforts. Discuss how individual and collective efforts to reduce food waste at school and home can support the achievement of the SDGs. This connection reinforces the idea that local actions have global impacts and empowers students to take meaningful steps toward sustainability.

By understanding the UN SDGs and their connection to food waste, students can appreciate the broader context of their actions and feel motivated to contribute to global sustainability efforts.

Connecting Food Waste to Broader Sustainability Issues

Reducing food waste is a critical component of addressing broader sustainability issues, such as resource conservation, climate change, and ethical food systems. By exploring these connections, students can develop a comprehensive understanding of the environmental, economic, and social impacts of food waste and the importance of adopting sustainable practices.

Resource Conservation:

Food waste represents a significant loss of resources, including water, energy, and land. Explain that the production, processing, and transportation of food require substantial amounts of these resources.

When food is wasted, the resources used to produce it are also wasted. Discuss the following points:

- Water Conservation: Agriculture accounts for about 70% of global freshwater use. Wasting food means wasting the water used to grow it. Highlight the importance of conserving water by reducing food waste and adopting efficient agricultural practices.

- Energy Use: Food production and distribution require energy for machinery, transportation, and storage. Reducing food waste decreases the demand for energy-intensive activities, helping to conserve energy and reduce greenhouse gas emissions.

- Land Use: Agriculture occupies a significant portion of the Earth's land surface. Wasting food contributes to deforestation and habitat loss, as more land is cleared for farming. Emphasize the importance of preserving natural habitats by minimizing food waste.

Climate Change:

Food waste contributes to climate change through the production of greenhouse gases. When organic waste decomposes in landfills, it produces methane, a greenhouse gas that is much more effective at trapping heat in the atmosphere than carbon dioxide. Discuss the following points:

- Reducing Emissions: Explain that reducing food waste can significantly lower greenhouse gas emissions. Highlight how composting organic waste, instead of sending it to landfills, can reduce methane production and enrich soil health.

- Sustainable Agriculture: Promote the adoption of sustainable agricultural practices that minimize food waste and reduce the environmental impact of farming. Discuss methods such as crop rotation, agroforestry, and integrated pest management that contribute to climate resilience and food security.

Ethical Food Systems:

Food waste raises ethical concerns related to food security and social justice. While millions of people around the world suffer from hunger and malnutrition, a significant portion of food is wasted. Discuss the following points:

- Food Security: Reducing food waste can improve food security by making more food available to those in need. Highlight initiatives that rescue surplus food and redistribute it to food banks and shelters, addressing hunger and reducing waste simultaneously.

- Equity and Justice: Explore the social implications of food waste and the importance of creating equitable food systems. Discuss how reducing waste can contribute to fairer distribution of resources and support marginalized communities.

Educational Activities:

Incorporate activities that help students understand the connections between food waste and broader sustainability issues. For example, organize a debate on the ethical implications of food waste, or create a project where students calculate the environmental footprint of their food choices and explore ways to reduce waste. Encourage students to think critically about the sustainability of their actions and the impact on future generations.

Empowering Action:

Encourage students to take action by adopting sustainable practices in their daily lives. Provide practical tips for reducing food waste, such as meal planning, proper food storage, and composting. Empower students to become advocates for sustainability by sharing their knowledge and inspiring others to make positive changes.

By connecting food waste to broader sustainability issues, students can develop a holistic understanding of the impacts of waste and the importance of sustainable practices. This comprehensive approach fosters a sense of responsibility and empowers students to contribute to a more sustainable and equitable world.

Teaching Sustainable Practices

Teaching sustainable practices is crucial for equipping students with the knowledge and skills they need to reduce food waste and promote sustainability. By focusing on practical actions such as reducing, reusing, and recycling, as well as making sustainable food choices, educators can empower students to make a positive impact on their environment and community.

Reducing, Reusing, and Recycling

The principles of reducing, reusing, and recycling are foundational to sustainable practices and are essential for minimizing food waste. By teaching these concepts, educators can help students develop habits that contribute to environmental sustainability.

Reducing Food Waste:

Reducing food waste begins with awareness and intentional actions to minimize the amount of food that goes uneaten. Educators can teach students about the importance of planning meals and making mindful purchasing decisions. Key strategies include:

- Meal Planning: Encourage students to plan their meals ahead of time to ensure they buy only what they need. This reduces the likelihood of food going bad before it can be eaten. Provide examples of meal plans and shopping lists to help students practice this skill.

- Portion Control: Teach students about appropriate portion sizes to avoid over-serving and subsequent waste. Use visual aids, such as portion size charts, to help students understand how much food they actually need.

- Mindful Purchasing: Discuss the importance of being mindful when shopping for groceries. Encourage students to check their pantry and refrigerator before shopping to avoid buying items they already have. Highlight the benefits of buying in bulk for non-perishable items and purchasing perishable goods in smaller quantities.

Reusing Leftovers:

Reusing leftovers is an effective way to reduce food waste and maximize the value of food. Educators can introduce creative ways to repurpose leftovers into new meals. Strategies include:

- Recipe Ideas: Provide students with recipes that use common leftovers. For example, leftover vegetables can be used in soups, stews, or stir-fries, while stale bread can be transformed into croutons or bread pudding.

- Storage Tips: Teach students proper storage techniques to extend the shelf life of leftovers. Discuss the use of airtight containers, refrigeration, and freezing to keep food fresh longer.

- Meal Prep: Encourage students to incorporate leftovers into their meal prep routines. Show how planning meals around leftovers can save time, reduce waste, and ensure that all food is used efficiently.

Recycling Food Waste:

Recycling food waste through composting is a sustainable practice that benefits the environment by returning nutrients to the soil and reducing landfill waste. Educators can provide hands-on opportunities for students to learn about composting. Key points include:

- Composting Basics: Introduce the basics of composting, including what can and cannot be composted. Discuss the difference between "green" materials (e.g., fruit and vegetable scraps) and "brown" materials (e.g., leaves, cardboard) and the importance of maintaining a balance.

- Composting Methods: Explain different composting methods, such as backyard composting, vermicomposting (using worms), and community composting programs. Provide examples of each method and discuss their benefits and challenges.

- Setting Up a Compost Bin: Guide students in setting up a compost bin at home or at school. Demonstrate how to layer materials, maintain the compost pile by turning it regularly, and monitor the decomposition process.

By teaching the principles of reducing, reusing, and recycling, educators can help students develop sustainable habits that minimize food waste and contribute to a healthier environment.

Sustainable Food Choices

Making sustainable food choices is another important aspect of reducing food waste and promoting environmental sustainability. Educators can guide students in understanding the impact of their food choices and encourage them to adopt more sustainable eating habits.

Understanding the Environmental Impact:

Educate students about the environmental impact of different types of food. Discuss how food production affects natural resources, such as water, land, and energy. Highlight the environmental footprint of various foods, including the emissions associated with their production, transportation, and disposal. Key points include:

- Plant-Based Diets: Explain that plant-based diets typically have a lower environmental impact compared to diets high in animal products. Discuss the benefits of incorporating more fruits, vegetables, legumes, and whole grains into their meals.

- Locally Sourced Food: Emphasize the benefits of buying locally sourced food, which often requires fewer resources for transportation and supports local farmers. Encourage students to visit farmers' markets and learn about seasonal produce.

- Organic and Sustainable Farming: Discuss the advantages of organic and sustainably farmed food, which avoids harmful pesticides and promotes biodiversity. Explain how these farming practices contribute to soil health and reduce pollution.

Choosing Sustainable Options:

Provide practical tips for making sustainable food choices. Help students understand how to evaluate the sustainability of their food and make informed decisions. Strategies include:

- Reading Labels: Teach students how to read food labels to identify sustainably sourced products. Look for certifications such as organic, Fair Trade, and Rainforest Alliance, which indicate sustainable practices.

- Reducing Meat Consumption: Encourage students to reduce their meat consumption by incorporating "Meatless Mondays" or other plant-based meal days into their routines. Provide recipes and ideas for delicious and nutritious plant-based meals.

- Minimizing Packaging Waste: Discuss the importance of choosing foods with minimal packaging to reduce waste. Encourage students to buy in bulk, use reusable produce bags, and avoid single-use plastics.

Promoting a Sustainable Food Culture:

Foster a culture of sustainability within the school and community by promoting sustainable food practices. Encourage students to share their knowledge and inspire others to make sustainable choices. Activities include:

- Sustainable Food Projects: Organize projects that involve researching and promoting sustainable food practices. For example, students can create presentations on the benefits of plant-based diets or develop a guide to local farmers' markets.

- Cooking Classes: Offer cooking classes that focus on preparing sustainable meals. Teach students how to cook with seasonal produce, minimize food waste, and create delicious plant-based dishes.

- School Garden Programs: Establish or enhance school garden programs to provide hands-on learning experiences. Students can grow their own fruits and vegetables, learn about sustainable gardening practices, and appreciate the effort required to produce food.

By teaching students to make sustainable food choices, educators can help them understand the broader impact of their eating habits and empower them to contribute to a more sustainable food system. These practices not only reduce food waste but also promote health, environmental stewardship, and social responsibility.

Long-Term Impact of Food Waste Education

Food waste education has the potential to create lasting impacts by instilling lifelong habits and fostering a culture of sustainability. By educating students about the importance of reducing food waste, schools can empower them to make sustainable choices throughout their lives and contribute to a healthier planet.

Creating Lifelong Habits

One of the most significant outcomes of food waste education is the development of lifelong habits that prioritize sustainability and environmental responsibility. By learning about the causes and impacts of food waste at an early age, students can adopt practices that reduce waste and conserve resources.

Early Education and Habit Formation:

Early education plays a crucial role in shaping behaviors and attitudes. When students are taught about food waste reduction and sustainability in their formative years, these concepts become ingrained in their daily routines. For example, students who learn to compost, plan meals, and store food properly are more likely to continue these practices into adulthood.

Practical Skills for Daily Life:

Food waste education provides students with practical skills that they can apply in their daily lives. Teaching students how to minimize food waste through meal planning, portion control, and creative use of leftovers equips them with valuable tools for managing their resources efficiently. These skills not only reduce waste but also promote financial savings and healthier eating habits.

Empowerment and Responsibility:

Educating students about food waste empowers them to take responsibility for their actions and make informed decisions. By understanding the broader impacts of their choices, students are motivated to adopt sustainable behaviors that benefit both the environment and society. This sense of responsibility extends beyond the classroom, influencing their actions at home, in their communities, and throughout their lives.

By fostering lifelong habits through food waste education, schools can create a generation of environmentally conscious individuals who are committed to sustainability.

Building a Culture of Sustainability

In addition to shaping individual behaviors, food waste education contributes to building a culture of sustainability within the school and the broader community. This cultural shift is essential for creating lasting change and promoting collective action toward environmental stewardship.

School-Wide Initiatives:

Implementing school-wide initiatives that focus on reducing food waste helps establish sustainability as a core value within the school community. Programs such as composting, food sharing, and waste audits engage students, staff, and parents in collective efforts to minimize waste. These initiatives demonstrate the school's commitment to sustainability and encourage everyone to participate in creating a waste-conscious environment.

Community Engagement:

Engaging the broader community in food waste reduction efforts amplifies the impact of educational initiatives. Schools can collaborate with local organizations, businesses, and government agencies to promote sustainable practices and share resources. Community events, workshops, and campaigns help raise awareness and foster a sense of collective responsibility for reducing food waste.

Role Models and Leadership:

Students who are educated about food waste and sustainability often become role models and leaders within their communities. By sharing their knowledge and advocating for sustainable practices, they inspire others to take action. This ripple effect helps spread the message of sustainability and encourages widespread adoption of waste reduction behaviors.

Long-Term Vision:

Building a culture of sustainability requires a long-term vision and commitment. Schools play a pivotal role in nurturing this culture by integrating sustainability into their curriculum, policies, and practices. By prioritizing food waste education and promoting sustainable behaviors, schools can contribute to a more sustainable future for all.

Through creating lifelong habits and building a culture of sustainability, food waste education has the power to drive meaningful and lasting change. This holistic approach ensures that students are not only equipped with the knowledge and skills to reduce waste but also inspired to lead and advocate for a more sustainable world.

Conclusion

The conclusion of this book aims to encapsulate the essential insights discussed and provide a roadmap for future efforts in food waste education. By summarizing the key points, exploring future directions, and issuing a call to action, this chapter seeks to empower educators, students, and communities to continue their efforts in reducing food waste and promoting sustainability.

Summary of Key Points

Throughout this book, we have explored the multifaceted issue of food waste and its far-reaching impacts. We began by understanding the definition and global scale of food waste, highlighting its environmental, economic, and social consequences. The exploration of food waste in the curriculum emphasized the importance of integrating this topic into various subjects, fostering a holistic understanding among students.

We delved into practical classroom activities, such as composting projects, growing food in the classroom, and utilizing digital resources and tools. These hands-on experiences not only engage students but also equip them with practical skills and knowledge to reduce food waste.

School-wide initiatives, including cafeteria waste audits and food sharing programs, were discussed as effective strategies for minimizing food waste at an institutional level. The involvement of students in campaigns and community engagement activities further demonstrated the importance of collective action in driving meaningful change.

Monitoring and evaluating food waste was underscored as a critical component of effective waste reduction efforts. Setting measurable goals, tracking progress, and analyzing data were highlighted as essential practices for ensuring the success and sustainability of food waste initiatives.

Finally, we examined the broader context of food waste and sustainability, linking food waste reduction to the United Nations Sustainable Development Goals (UN SDGs) and broader environmental issues. Teaching sustainable practices and fostering lifelong habits were presented as key strategies for creating a culture of sustainability within schools and communities.

By summarizing these key points, we can appreciate the comprehensive approach required to address food waste effectively. Each chapter has provided valuable insights and practical strategies that educators can implement to foster a sustainable future.

Future Directions in Food Waste Education

As we look to the future, emerging trends and technologies offer new opportunities for enhancing food waste education and reduction efforts. Advances in technology, such as artificial intelligence (AI) and machine learning, can provide innovative solutions for monitoring and managing food waste. For example, AI-powered tools can analyze food waste data in real-time, offering insights and recommendations for reducing waste more effectively.

Technological innovations in food production, storage, and transportation also hold promise for minimizing food waste. Smart packaging, for instance, can monitor the freshness of food and provide alerts when items are nearing their expiration date. Such technologies can help consumers make informed decisions and reduce the likelihood of food spoilage.

Furthermore, digital platforms and apps can facilitate better food management practices. Apps that track food inventory, suggest recipes based on available ingredients, and connect users with local food donation programs can empower individuals to reduce waste at home and in their communities.

Policy and institutional support play a crucial role in driving systemic change and promoting food waste reduction at a broader scale. Governments and educational institutions must prioritize policies that encourage sustainable practices and provide the necessary resources and infrastructure.

At the policy level, initiatives such as mandatory food waste audits, incentives for food donation, and regulations on food labeling can create an environment conducive to waste reduction. Schools can advocate for and implement policies that support sustainable practices, such as composting programs, zero-waste initiatives, and sustainable procurement policies.

Institutional support is also vital for sustaining food waste reduction efforts. Educational institutions should invest in training and professional development for teachers, providing them with the knowledge and tools to effectively integrate food waste education into their curricula. Collaboration with local organizations, businesses, and government agencies can further strengthen these efforts and create a supportive network for sustainability initiatives.

By embracing emerging trends and securing policy and institutional support, we can enhance the effectiveness of food waste education and create a lasting impact.

Call to Action

Teachers play a pivotal role in shaping the attitudes and behaviors of future generations. As educators, it is crucial to take initiative in integrating food waste education into the curriculum and fostering a culture of sustainability within the classroom. By actively seeking out resources, collaborating with colleagues, and participating in professional development opportunities, teachers can stay informed about best practices and innovative strategies for reducing food waste.

Encourage teachers to create engaging and interactive lessons that highlight the importance of food waste reduction and sustainable practices. Incorporate hands-on activities, such as composting projects and food audits, to provide students with practical experiences that reinforce their learning. Utilize digital tools and resources to enhance the educational experience and make it relevant to students' everyday lives.

Moreover, teachers should serve as role models by demonstrating sustainable behaviors and practices. Simple actions, such as reducing personal food waste, promoting waste-free lunch initiatives, and advocating for environmentally friendly practices within the school, can inspire students to adopt similar habits.

Addressing food waste requires a collective effort that extends beyond the classroom. Schools, families, and communities must work together to create a sustainable environment where food waste reduction is a shared priority. Building a collective effort involves fostering collaboration, raising awareness, and promoting community engagement.

Schools can organize events, workshops, and campaigns that involve students, parents, and community members in food waste reduction initiatives. These activities provide opportunities for learning, collaboration, and action, reinforcing the importance of collective responsibility. Engage local organizations, businesses, and government agencies to support and amplify these efforts, creating a network of stakeholders committed to sustainability.

Families play a crucial role in reinforcing the lessons learned at school. Encourage parents to participate in school activities, adopt sustainable practices at home, and support their children's efforts to reduce food waste. Providing resources and information to families can help them understand the significance of food waste and empower them to make positive changes.

By building a collective effort, we can create a culture of sustainability that extends beyond the school walls and into the broader community. Together, we can make a significant impact on reducing food waste and promoting environmental stewardship.

In conclusion, the fight against food waste is a multifaceted challenge that requires a comprehensive and collaborative approach. By summarizing the key points, exploring future directions, and issuing a call to action, this chapter aims to inspire and empower educators, students, and communities to take meaningful steps toward reducing food waste and fostering a sustainable future.

www.ingramcontent.com/pod-product-compliance
Lightning Source LLC
Chambersburg PA
CBHW071950210526
45479CB00003B/886